Published by Smart Apple Media
1980 Lookout Drive, North Mankato, MN 56003

Designed by Stephanie Blumenthal
Production Design by Kathy Petelinsek

Photographs by Maslowski Wildlife Photography

Copyright © 2002 Smart Apple Media.
International copyrights reserved in all countries.
No part of this book may be reproduced in any form without
written permission from the publisher.

Library of Congress Cataloging-in-Publication Data

Maslowski, Stephen. Birds in winter / by Steve Maslowski with Adele Richardson.
p. cm. — (Through the seasons ; 4)
Summary: Focuses on the activities of various kinds of birds
which remain in the northern areas during the winter including their
search for food and use of feathers for warmth.
ISBN 1-58340-059-1
1. Birds—Wintering—Juvenile literature. [1. Birds] I. Richardson, Adele. II. Title.

QL698.3 .M36	2001
598.156—dc21	99-046946

First Edition

2 4 6 8 9 7 5 3 1

THROUGH THE SEASONS

BIRDS IN WINTER

**Text by Steve Maslowski with Adele Richardson
Photographs by Maslowski Wildlife Photography**

SMART APPLE MEDIA

The coming of winter changes much of the land across North America. Many trees drop their colorful autumn leaves. The sun shines less and less until nights last as long as 14 hours. Finally, much of the northern half of the continent is covered by a thick blanket of snow. The season of winter starts on the 21st day of December, bringing temperatures that can dip far below freezing in many places.

Water freezes when the temperature reaches 32° F (0° C).

Many birds fly south before winter sets in. There they will find warmer temperatures and more food. Not all birds leave, though. Some live year-round in the cold regions of the upper United States and Canada. These birds must search for food every day in order to survive.

Many birds that spend the winter in the North are quite small—less than seven inches (18 cm) long.

Flock of Canada geese

They are able to survive the cold only by eating a lot of food. The energy they get from eating helps them stay warm. These birds have huge appetites, eating as much as one-third of their body weight each day. Most small birds would not survive a single day in winter if they didn't eat.

Food can be difficult to find during this season. Some nuthatches, cardinals, and robins may spend the

> *A bird's body temperature is very warm— usually between 106° and 113° F (41° to 45° C).*

Goldfinch (top); robin (bottom)

winter as far north as Canada. They eat insects and insect eggs found in trees and bushes. Any fruit and berries that the birds can find are quickly gobbled down as well. An especially tasty winter food for robins is hawthorn fruit, which is like tiny apples.

Birds eat plant seeds when they can be found. Seeds may be left

The snowy owl will live in cold regions all year as long as it can find food. This owl grows special feathers on its face during the winter. Tiny feathers cover the front of the bird's nostrils. This keeps cold, blowing snow out but still lets air in so that the owl can breathe.

Quail eating seeds

over in fields after farmers have harvested crops. Others are spilled around grain elevators where seed is stored throughout the winter months. Meadows and swamps may also have a few weeds that have produced seeds.

Some birds find winter food in feeders that people hang in their backyards. Sunflower seeds, millet, peanuts, and cracked corn are common types of store-bought birdseed. Some types of feeders may also hold suet, which is hard, raw beef fat. Suet is a good winter food because it contains nutrients that give birds the energy they need to keep warm. Birds that like suet include woodpeckers and chickadees.

Flock of turkeys

Whether a feeder has seeds or suet inside, once it is placed out for winter birds, it must stay stocked until spring. This is important because the birds come to rely on the food being there. If it is suddenly taken away, the birds may not be able to find another sufficient food supply nearby and may die.

The hairy woodpecker lives in the North during the winter, eating insects found under tree bark. To get to this food, the bird hammers a small hole in the tree with its strong, thick bill. Then it sticks its long tongue inside to grab the insects. The hairy woodpecker helps trees by eating insects that can damage the bark.

Many larger birds are predators—animals that eat other animals for food. Snowy owls, eagles, and falcons are all predators. Many live in wooded areas in the northern United States and Canada. As long as there is food to be found, predatory birds may live in these cold regions all year long. Much of their winter diet consists of

The nuthatch is a small bird that lives in northern wooded areas. It ranges in size from seven to nine inches (18–23 cm) long. Nuthatches are often seen hopping down tree trunks upside down while searching for food. They eat mainly insects found in the cracks of tree bark.

Black-capped chickadee eating suet

rodents, such as lemmings, mice, and small rabbits. Eagles may also pluck fish out of lakes that have not frozen over.

Predatory birds use their great vision and hearing to hunt for food. Because the sun does not shine long during winter days, predators can't always rely on

Owl looking for prey

The blue jay spends the winter amid the snow of Canada and the upper United States. It is about 12 inches (25 cm) long and is blue and white with beautiful black stripes. Blue jays are common visitors to bird feeders, where they often draw attention to themselves with their loud calls.

Snowy owl

their eyesight. Many times, the birds must glide through the darkened sky listening for food. Their ears are so sharp that they can even hear rodents trying to hide under the snow. Once the birds have pinpointed the location of their prey, they quietly swoop down to snatch it with their sharp claws.

Although birds would not last long in winter without food, they can go a long time without drinking water. This is good, since most ponds and streams are frozen over during this time of year. Any birds that eat fruit, insects, or other animals get most of the water they need from their food.

Food is only part of what keeps birds warm in winter. They also rely on their feathers. A bird's feathers are like its clothing—they keep heat in and cold out. Just as people wear more clothes in winter, birds also wear more feathers. The complete covering of feathers on a bird's body is called its plumage.

Bobwhites huddled together

Although a bird's feathers are very light, they keep its body warm and dry. Each feather has a shaft that is stiff and strong, like the trunk of a tree. Sticking out from this shaft are barbs—tiny hairs that are similar to the branches of a tree. Sticking out from the barbs are even smaller hairs called barbules.

The brightly-colored cardinal lives all over North America east of the Rocky Mountains. It does not migrate south; instead, it stays in the area where it will build a nest in the coming spring. Cardinals may live as far north as Canada or as far south as Florida.

Purple finch

Many types of finches, such as the northern junco and the house finch, live in the North during the winter. These small birds usually eat only seeds and can often be seen around winter feeders, where they gather every day to consume large amounts of food.

At the end of each barbule is a tiny hook that connects to the other barbules' hooks. This keeps the feathers linked tightly together so that cold air and rain can not get through to the bird's body. The feathers partly overlap one another, much like shingles on a roof. If a gust of wind blows some feathers out of place, the bird shakes itself.

This makes the feathers fall back into place, and the hooks reconnect.

To help keep their feathers smooth and straight, birds run their beak over them. This method of combing, called preening, also helps to spread around an oil that keeps the feathers waterproof. The oil comes from a gland on birds' backs. Birds release a small amount of the oil by pressing gently on the gland with their beak.

Under the oiled and combed feathers is another layer of soft, fluffy feathers called down. When the temperature gets very cold, a bird will fluff up its down feathers. This lets in a little air, which is then warmed by the bird's body heat. The warmed air acts like a blanket next to the bird's skin.

In the northernmost part of Canada, winter temperatures often dip as low as 80° F below zero (−62° C).

White-tailed ptarmigan

Birds do not grow feathers on their feet, which don't need to stay warm like the body does. Birds' feet are mostly just bone and scaly skin and can be exposed to cold water, snow, or air for a long time. If their feet do get cold, birds can fluff their feathers down over them for warmth.

> Snow is frozen water that has formed into tiny ice crystals in the atmosphere.

The season of winter lasts until the 20th day of March. Most birds will survive even the coldest winter as long as they are able to find enough food. As spring approaches, the days become longer and food becomes easier to find. Soon, the birds that flew south will return home to join those that stayed in the North.

Wild turkeys

INDEX

bird feeders 8, 10, 12, 19
    suet 8, 10–11
blue jays 9, 12
bobwhites 16–17
Canada geese 4
cardinals 5, 18
chickadees 8, 10–11
    black-capped 10–11
eagles 11, 12
falcons 11
feathers 7, 16, 18–20, 23
    down 20
    preening 20
    structure 18–19
feeding 4, 5, 7–8, 10, 11–12, 16, 19
finches 5, 18–19
    goldfinches 5
    purple 18–19

migration 4, 23
nuthatches 5, 11
owls 7, 11, 12, 13
    snowy 7, 11, 13
predatory birds 11–13
    hearing 13
    hunting 12–13
quail 6–7
robins 5, 7
temperatures 4, 5, 20
    body 5
turkeys 8, 23
white-tailed ptarmigans 20–21
woodpeckers 8, 10
    hairy 10